EXTREME WORLD
THE WORLD'S MOST FANTASTIC PLANTS

by Cari Meister

PEBBLE
a capstone imprint

Published by Pebble, an imprint of Capstone.
1710 Roe Crest Drive
North Mankato, Minnesota 56003
capstonepub.com

Copyright © 2023 by Capstone. All rights reserved. No part of this publication may be reproduced in whole or in part, or stored in a retrieval system, or transmitted in any form or by any means, electronic, mechanical, photocopying, recording, or otherwise, without written permission of the publisher.

Library of Congress Cataloging-in-Publication Data is available on the Library of Congress website

ISBN: 9781666348385 (hardcover)
ISBN: 9781666348422 (paperback)
ISBN: 9781666348460 (ebook PDF)

Summary: Presents the most incredible plants on Earth, including some of the tallest, smelliest, thorniest, and more.

Editorial Credits
Editor: Christopher Harbo; Designer: Kay Fraser; Media Researcher: Svetlana Zhurkin; Production Specialist: Katy LaVigne

Image Credits
Shutterstock: ChameleonsEye, 12, Diana Taliun, 20 (bottom left), Donquixote Rino, 7, guentermanaus, 14, Heather Lucia Snow, 10, Herrieynaha, 6, I'm friday, 20 (top right), Joy Brown, 9, Juli V, 15, Kuttelvaserova Stuchelova, 13, Lester Balajadia, 19, Lippert Photography, 11, Lucky-photographer, 4, Mazur Travel, cover, 1, Ongushi, 16, SantaLiza, 5, SKphotographer, 17, Stephen Moehle, 18, Virtexie, 20 (bottom right)

All internet sites appearing in back matter were available and accurate when this book was sent to press.

Printed and bound in the USA. 4882

TABLE OF CONTENTS

Plants Are Fantastic! 4

Huge Smelly Flowers 6

Plump Pumpkins 8

Fire Starters 10

Deadly Trappers 12

Thorny Trees 14

Fast Growers 16

Living Skyscrapers 18

    Fantastic Fairy Garden 20

    Glossary 22

    Read More 23

    Internet Sites 23

    About the Author 24

    Index 24

    Words in **bold** are in the glossary.

PLANTS ARE FANTASTIC!

Plants grow all over the world. Giant trees stretch toward the sky. Bright flowers blanket mountain meadows. Prickly **cactus** grow in dry deserts.

Plants can be very different, but they all have something in common. All plants need sunlight and water to grow. Take a look at some of the most fantastic plants on Earth!

HUGE SMELLY FLOWERS

Two of the largest flowers grow in rain forests. The *Rafflesia arnoldii* is a bright red flower. It can grow more than 3 feet (0.9 meter) across.

The corpse flower is even bigger. It can grow up to 10 feet (3 m) tall.

Besides being huge, both flowers are very stinky. They smell like rotting meat!

PLUMP PUMPKINS

Lucky for us, many plants grow tasty fruits and vegetables. From pies to salads, plants are part of our everyday eating. So what is the biggest fruit of all? It's the pumpkin. Some pumpkins can weigh more than 2,000 pounds (907 kilograms)!

FIRE STARTERS

Some plants like it hot. Eucalyptus trees have **oils** in their bark and leaves that can catch fire! But fire isn't all bad. Heat from a fire causes their **seed pods** to open and release the seeds. When they do, new eucalyptus trees can begin to grow.

eucalyptus seed pods

DEADLY TRAPPERS

Look out, it's a trap! A fly sees some sticky nectar. It lands on the open leaf of a Venus flytrap and starts to slurp. The fly's leg hits a tall, thin **trigger hair**.

Nothing happens at first. But when the fly's leg hits the hair again. Snap! The leaf slams shut. The fly tries to get out. But it is stuck in a meat-eating plant!

THORNY TREES

Long ago, **mammoths** roamed the Earth. They ate the leaves, bark, and sweet seed pods from the honey locust tree. Over time, the trees **adapted**. They grew thorns to keep animals from eating them. Today, their thorns are very long and sharp.

honey locust seed pods

honey locust thorns

FAST GROWERS

Duckweed floats in many ponds and lakes. It may be tiny, but this seed-like plant grows faster than any other plant. A patch of duckweed can double in size in one day!

Some types of duckweed are healthy to eat. People call the tiny plants "water eggs." It tastes similar to sweet cabbage.

LIVING SKYSCRAPERS

Redwoods are nature's **skyscrapers**. These giant trees tower over the ground. The tallest redwood is named Hyperion. It is 380.8 feet (116.1 m) tall!

Hyperion

About 400,000 different kinds of plants live on Earth. They feed us and **shelter** us. The more we learn about them, the more we are amazed. Long live plants!

FANTASTIC FAIRY GARDEN

Create your own little fairy garden with grass seed and natural materials you can find outside. Depending on what you use, your garden can have a woodsy, beach, or city theme!

WHAT YOU NEED
- a shallow pot or large bowl
- potting soil
- natural materials, such as sand, shells, pebbles, twigs, and moss
- grass seed
- water

WHAT YOU DO

1. Fill a shallow pot or large bowl with potting soil.

2. Use natural materials to make a little home or a garden for a fairy. Use sand, shells, or pebbles to make a path. Make a tiny house with twigs and moss. Be creative!

3. Sprinkle grass seed anywhere in your fairy garden that you would like grass to grow. You can sprinkle it all over, along a path, or in front of a little house to make a tiny lawn.

4. Place your fairy garden outside or in a sunny window.

5. Water the grass seed in your fairy garden every day. Watch as the grass sprouts and grows each day!

GLOSSARY

adapt (uh-DAPT)—to change to fit into a new or different environment

cactus (KACK-tus)—a plant covered in spines that is found in desert areas

mammoth (MAM-uhth)—an extinct animal like a large elephant with long, curved tusks and shaggy hair

oil (OIL)—a slippery liquid that does not mix with water

seed pod (SEED POD)—a container that a plant makes to hold its seeds

shelter (SHEL-tur)—to protect

skyscraper (SKYE-skray-pur)—a very tall building made of steel, concrete, and glass

trigger hair (TRIG-uhr HARE)—a short, stiff hair of a plant that senses when things touch it

READ MORE

Fox, Kate Allen. *Pando: The True Story of a Living Giant.* North Mankato, MN: Capstone, 2021.

Hirsch, Rebecca E. *When Plants Attack: Strange and Terrifying Plants.* Minneapolis: Millbrook Press, 2019.

Markovics, Joyce L. *Spiky Plants.* Ann Arbor, MI: Cherry Lake Press, 2021.

INTERNET SITES

Meat Eating Plants
kids.nationalgeographic.com/science/article/meat-eating-plants

Plant Facts
dkfindout.com/us/animals-and-nature/plants

Science4Fun: Plants
science4fun.info/plants

ABOUT THE AUTHOR

Cari Meister has written more than 130 books for children, including the Tiny series (Penguin) and the Fast Forward Fairy Tales series (Scholastic). Cari is a school librarian who loves to visit other schools to talk about the joy of reading and writing. Cari lives in the mountains of Colorado with her husband, four boys, one horse, and one dog. You can find out more about her at www.carimeister.com.

INDEX

cactus, 4

flowers, 4, 6–7

fruits, 8

meat-eating plants, 12

seed pods, 10, 14

sunlight, 5

thorns, 14

trees, 4, 10, 14, 18

vegetables, 8, 17

water plants, 16–17